Tackling Under-Development in the Niger-Delta through a Dynamic Approach

Ovuokeroye Edih

Bibliographic information published by the German National Library:

The German National Library lists this publication in the National Bibliography; detailed bibliographic data are available on the Internet at http://dnb.dnb.de.

ISBN: 9783346580788
This book is also available as an ebook.

Print and binding: Books on Demand GmbH, Norderstedt, Germany
Printed on acid-free paper from responsible sources.

GRIN web shop: https://www.grin.com/document/1165757

TACKLING UNDER-DEVELOPMENT IN THE NIGER-DELTA THROUGH A DYNAMIC APPROACH

EDIH UNIVERSITYOVUOKEROYE

Being a Dissertation submitted to Post Graduate School in partial fulfillment of the requirements for the award of Masters of Science (M.Sc) degree in Business Management, Department of Business Administration, Delta State University, Abraka, Nigeria.

March, 2016

Table of Contents

ABSTRACT

The study analyzed a mathematical model that would help to facilitate the development of the Niger Delta region in a phase by phase approach. The region has been bedeviled with the problem of underdevelopment culminating into youth restiveness, violence and militancy, poverty and environmental degradation despite the abundant oil wealth. The study utilized secondary data. A network model was drawn to practically and mathematically demonstrate the journey from under-development towards development and analyzed by applying the forward recursive equation formulated for study. Findings showed that the Niger Delta region could be developed through a phase by phase development plan. The study recommended that Federal and State governments, Ministry of Niger-Delta, Niger-Delta Development Commission (NDCC) and Multinational Companies are advised to be very sincere to the development plan and also adopt a dynamic approach in executing development projects and programmes for the region.

Key words: Underdevelopment, Development, Dynamic approach, Mathematical model and Niger-Delta.

CHAPTER ONE

INTRODUCTION

1.1Background to the Study

The Niger Delta region has been begging for development since independence in 1960.The region was adjudged the producer of oil and gas since 1956 in commercial quantity which contributes over 90% of national income of the country. However, the region has been bedeviled by underdevelopment due to neglect by successive governments of the country till date. As a result of the deliberate abandonment and insensitivity of the government to the peculiar characteristics of the Niger- Delta region, unwarranted upheavals have brewed in the region.Currently, it has gone out of proportion and poses an uphill task to the Federal and State governments.

Though, several ad hoc commissions and summits had been set up to look into the unique nature of the Niger-Delta question, no concrete development has been recorded in the area. There has been the Sir Henry Willinks Commission in 1957 in response to the concern of ethnic minorities over their perceived slim chance of survival in the Nigerian enterprise by colonial administration. There had been the Niger Delta Development Board (NDDB) via Supplement Federal Government Gazette No56 vol. 46 of Sept. 1959, Niger-Delta Basin Development Authority(NDBDA) Decree 1979, Special Fund Oil Producing Area by Revenue Act of 1981, Presidential Task Force for the Development of Oil Producing Areas of 1989, Oil Mineral Producing Areas Development Committee (OMPADEC) of 1992, and subsisting Niger- Delta Development Commission,(NDDC), 2000, Gen Ogomudia Committee 2002, Coastal States, 2007, Technical Committee on Niger Delta of 2008, Creation of the Ministry of Niger-Delta in 2008, and Petroleum Industry Bill granting 10%

right to Oil bearing/impacting Communities in October, 2009 that was not implemented (Emma Okah, Sunday Sun, May 23, 2010).

From the foregoing, the Federal Government has made twelve different attempts to marshal a way for developing the region by successive military and civilian regimes. The question is what is actually hindering these committees from making appreciable progress? Without much ado, we can confidently say that, there is lack of political will to stamp a concrete and comprehensive programme for the real development of the region. These frantic efforts of government were mere political charades in the words of former Vice-President of Nigeria, Dr. Alex Ekwueme and open avenues to perfect their tricks. Gen. Useni argued that, solving the problem in the Niger-Delta goes beyond the establishment of a ministry. He advocated that, leadership must be sincere, fair and just in getting the right people to solve the problem (Vanguard Tuesday, October 28, 2008).

The Niger- Delta environment is depleting every day as a result of the oil and gas activities that are destroying the eco-system which is posing huge challenges to the Government and the well-being of the people. In similar vein, the President of Nigerian Conservative Foundation (NCF) noted that, within the context of these challenges, there are roles for the governments, individuals, Oil and Gas companies. He contended that, the government and multinationals should play leading roles and develop strategies for economic development with clear environmental agenda. A rapid and well planned strategy to reduce poverty, clean up the rivers and creeks, well laid out settlement with adequate infrastructure of energy, portable water supply, transportation are some of the ways to resolving the grave crises in the Niger-Delta (Financial Vanguard, Nov. 3, 2008).

1.2 Statement of the Problem

The problem of under-development in the Niger-Delta is no longer a mean issue because it has transcended from national to international dimension. It is complex and hydraheadedsince it has resulted to youth restiveness and taking of arms against the government. Jobless youths are tired of failed promises to provide job opportunities and enabling environment for private businesses. More so, the activities of oil exploration and exploitation have depleted the ecosystem. Environmental degradation through theactions of oil spills aredestroyingfarm lands and rivers.In such a precarious situation in the Niger- Delta, Mere political pronouncements or jamborees cannot and will not solve this sickness from its roots. This is why a multi- dimensional approach must be applied (if and onlyif) the government must develop the Niger-Delta area. It is against this backdrop that, the researcher is working on the possibility of developing adynamic plan for solving the problem of under-development plaguing the Niger- Delta region.

1.3 Objectives of the Study

The major objective of the study is to develop a model that will assist in solving the problem of under-development in the Niger-Delta. Consequently, the following specific objectives are to demonstrate that:

i. the use of an integrated development plan based on dynamic approach can solve the problem of underdevelopment.

ii. through a recursive equation underdevelopment issues can be solved by a network model.

1.4 Research Question.

The research question, can the problem of under-development be solved through an integrated development plan based on dynamic approach was examined

1.5 Research Hypothesis

There is no significant relationship between an integrated plan and development of the Niger-Delta region.

1.6 Scope of the Study

The study covers the Niger-Delta region in Nigeria. However, data collection was restricted to three States, namely, Delta, Bayelsa and Rivers and Niger-Delta Development

Commission (NDDC). The study considered 1967 - 2015 as the duration for assessing the impact of underdevelopment in the region.

1.7 Significance of the Study.

First, the dynamic approach (model) would help to address the problem of underdevelopment by simplifying it into solvable stages.

Second, stakeholders, government in particular would have a scientific blue print known as development plan to work on.

Lastly, if the recommendations of the study are diligently carried out by government and Multinational corporations, peace and security would return to the region and the country.

1.8 Limitations to the Study

Operation research being a new field of study in higher institutions in Nigeria, has not enough materials to consults as regards the peculiar problem facing the Niger-Delta. Another limitation was the difficulty faced with the conversion of project cost and timeframe of completed projects into numerals to ease the mathematical analysis of the model. Also, developing an appropriate recursive equation for the problem was a herculean task. However, through the rigorous and indepth study of related literature with specialty in diverse empirical studies these itemized limitations were surmounted.

1.9 Definition of Terms

Dynamic approach : This Concept entails a scientific and mathematical approach of decomposing a complex problem into phases to ease the process of arriving at a solution for the original issue.

Underdevelopment : This is state of affairs characterised with incidences of poverty, ignorance, or diseases occasioned by misdistribution of national income , administrative incompetence and corruption in public offices.

Development : This is situation that connotes an entire transformation of an economy from a less desirable to a more desirable state. It reflects improvement in the general well being of the citizens of a country.

Mathematical approach : It the procedure that uses the tools of additions, substractions, divisions and multiplications and formulas consciously built in a model for solving complex problems

CHAPTER TWO

REVIEW OF RELATED LITERATURE

2.1 Conceptual Review

The study reviewed related literature on the following sub-headings; development, underdevelopment, strategic planning, mathematical and economic model and dynamic programming. Empirical studies and theoretical framework on which the research was built were equally diagnosed.

Meaning of Development.

Development is a dynamic process. It is a many sided process in human society. At the level of the individual, it implies increased skill and capacity, greater freedom, creativity, self-discipline, responsibility and material well-being. Some of these are virtually moral categories and are difficult to evaluate depending as they do on the age, in which one lives, ones class origins, and one's personal code of what is right and what is wrong? However, what is indisputable is that the achievement of any of those aspects of personal development is dependent on the state of the society (Rodney, 1972).

Sanusi Daggash (Vanguard 14th October, 2008) opined that, development is a highly normative concept, which cannot be given a single line definition. He stressed that, it is a situation that is characterized by growth, progress, modernization advancement, and social transformation.

He further enunciated that, development is a process of transformation that brings about positive outcomes. It is multi-dimensional process involving structural, institutional, social, political growths and economic transformation. Technically, it's increased economic efficiency of a nation expansion in productivity on sustainable basis that adapts to various exogenous and endogenous shocks. He supports the views popularized by Amartya (1999) which focus on freedom. These include freedom from famine and malnutrition, freedom from poverty, access to healthcare and freedom from premature mortality.

Development to the layman is about consistent and sustainable improvement in the quality of his life as well as modernization of his environment. In fact, nations are categorized based on their level of development and prosperity, using indices such as income per capita.

McKinsley's Global Industries ranking of World Economics divided the world into three segments, namely , the foot hill countries' (with per capita income less than $7, 500), the 'slope countries' with per capita income between $7, 500 and $27, 000 and the 'mountain country' with per capita income above $27, 000. This categorization implies that development has as inverse relationship with poverty and quality of life. In fact, they are diametrically opposed to each other, as strange bed fellows (Vanguard Oct. 14, 2008). Nigeria per capita income is currently hovering around $1200 using Mckinsley's Global Ranking. This means, poverty is draining the population to death.

A number of economists have indeed emphasized the definition of development in terms of an increased in per capita income. Meier (1976), defines economic development as the process whereby the real per capita income of a country increases over a long period of time subject to the stipulations that the number below an absolute poverty line does not increase and the distribution of income does not become more unequal. Baran (cited in Jhingan, 2008) offers the view that economic development refers to as increase over-time in per capital output of material goods.

Economic development is nowadays seen as an end in the whole efforts to liberate the poverty—ridden people of the less developed countries (LDCs) not only from the shackles of ignorance, diseases, and squalor, but from the even more tasking problem of environmental degradation, political instability, threat to individual liberty, and unhealthy socio-cultural institution to changes in the faces of modem demands of economic development (Leonard 2000). This is exactly the case of the Niger Delta people. That is why the uproar is high.

It is therefore necessary that an acceptable definition of development must emphasize its multidimensional process involving changes in structures, attitudes and institutions as well as the acceleration of economic growth, the reduction of inequality and eradication of absolute poverty. This process enhances the worth of values of economic development to the countries that have achieved and can sustain it (Leonard, 2000).

Meaning of Underdevelopment

Under-development may not exactlymean the opposite of development. It is difficult to give a price criterion of underdevelopment. Underdevelopment can be defined in many ways, by the incidence of poverty, ignorance, or disease, by misdistribution of national income, by administrative incompetence and by social disorganization (Jhinghan, 2008).

The concept of 'underdevelopment' refers to an all-embracing condition of society, its social institutions, its political organization, its economic characteristic of the underdeveloped country. It is the economic ones that are the most readily listed and least likely to be disputed. They include, such phenomenon as the co-existence of market and predominantly self-subsistence economies (the former responding relatively quickly and the latter relatively slowly to financial inducements), the under-utilization of resources (capital, labour, land and material) and the absence of corrective institutions (Labour, exchange, bonds and shares market, discounting houses, future markets, marketing services, agricultural extension etc), the paucity of research, and the multiplicity of barriers to innovation, the adoption of techniques unsuited to the resource mix of the country, and the general failure to synchronize interdependent activities (Griffin 2004).

Underdevelopment is shocking, the squalor disease, unnecessary deaths and hopelessness of it all. No man understands if underdevelopment remains for him a mere statistics reflecting low income, poor housing, and premature mortality. The most emphatic observer can speak objectively about underdevelopment only after undergoing, personally or vicariously, the shock of underdevelopment". This is a unique culture shock to one who is initiated to the emotions, which prevail in the culture of poverty. The reverse shock is felt by those living in destitution when a new self- understanding reveals to them that their life is neither human nor inevitable(Goulet, 1971 & Leonard, 2000).

The prevalent emotion of under-development is a sense of personal and societal importance in the face of disease and death, of confusion and decisions govern the course of events, of hopelessness before hunger, and one cannot understand how cruel that hell is merely by gazing upon poverty as an object (Goulet, 1971).

Obviously, underdevelopment is not absence of development, because every people have developed in one way or another and to a greater or lesser extent. Underdevelopment makes sense only as means of comparing levels of development. It is very much tied to the face that human social development has been uneven and from a strictly economic view-point, some human groups have advanced further by producing more and becoming wealthier (Rodney, 1972).

He further argued that, at all times, the idea behind underdevelopment is a comparative one. It is possible to compare the country and determine whether or not it had developed, and more importantly, it is possible to compare the economies of any two countries or sets of countries at any given period in time. To buttress his point, Rodney pointed that a second and even more indispensable component of modern underdeveloped is that, it expresses a particular relationship of exploitation, namely, the exploitation of one country by another. All of the countries named as underdeveloped in the world are exploited by others, and the underdevelopment with which the world is now preoccupied is a product of capitalist, imperialist, and colonialist exploitation.

This is the exact case of the Niger-Delta region. The Federal Government and Oil companiesare milking the region of her treasured resource (crude oil) without commensurate developmental projects. With the available mineral and human resources in the Niger- Delta region, the region is bound to develop in leaps and bounds if other things being equal.

Strategic Planning Process

Strategic planning defines objectives and assesses both the internal and external situation to formulate a plan, implement the strategy, evaluate the progress, and make adjustment as necessary to stay on track. A strategic plan refers to the overall direction a business (or the development of State in the Niger- Delta region) will take over a longer term. Within the overall strategy, a business will have shorter terms; financial goals, marketing goals, production goals, and human resources goals.

The Niger- Delta region's development plan is never different from this illustration of a strategic business plan. There is the overall objective of developing the region to a desirable development standard but this cannot be achieved at one go. That was why the dynamic programming approach is being considered in the study as a way of phasing the development agenda for the region. A simplified view of the strategic planning process is shown by the following illustration:.

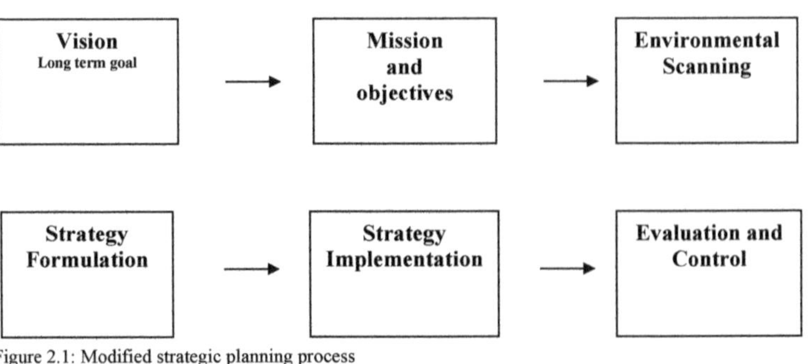

Figure 2.1: Modified strategic planning process
Source: goggle internet, www.strategic planning.

Mathematical or Economic Model

An economic model is an organized set of relationship that describes the functioning of an economic identity under a set of assumptions from which a conclusion or a set of conclusions is logically derived. The economic identity may be a household, a single industry, a region, an economy or the world as a whole. The Niger Delta is a region in Nigeria. As a matter of fact, economic model is a set of economic relationship which is generally expressed through a set of mathematical equations (Jhingan, 2008).

Limitations to Economic/Mathematical Models

Economic Models are beset with a number of limitations. Pure theoretical models do not provide full explanations or correct predications of the phenomenon under study.

a). Economic models are not comprehensive but partial due to uncontrollable variables.

b) When applied to real economic situations, they are selective, abstract and arbitrary. -

c). There are four ways in which errors enter into economic models as a result of assumptions which are not explicit.

i Certain parameters may stay constant.

ii. The number of strategic variables may be narrowed down by a single one.

iii. Dissimilar items may be analyzed.

iv Certain sequences may be isolated and analyzed without regards to their relationships to other sequences.

The short-comings notwithstanding, the importance of a model in simplifying thorny and abstract issues cannot be over-emphasized. It is therefore, suggested that, similar variables that are relevant to the problem to be analyzed and avoid ambiguity.

Meaning of Dynamic programming.

Dynamic programming (DP) determines the optimum solution of a multivariable problem by decomposing it into stages, each stage comprising a single variable sub problem. The advantage of the decomposition is that the optimization process at each stage involves one variable only, a simpler task computationally than dealing with all the variables simultaneously (Taha, 2007). The mathematical technique of optimizing sequence of interrelated decision over a period of time is called dynamic programming. It uses the idea of recursion to solve a complex problem, broken into a series of interrelated (sequential) decision stages (also called sub- problems) where the outcome of a decision at one stage affects the decision at each of next stage (Sharma, 2005).

Vohra (2007) defines dynamic programming as a useful quantitative analysis technique that can be use to solve many optimization problems. It deals with relatively large and complex problems, and involving making a sequence of interrelated decisions. The techniques provide a system procedure for determining optimal combination of decisions. A given problem may be complex if it is not possible to solve it in one go. It may be solved by breaking it into a series of smaller and more tractable sub-problems. This is exactly what dynamic programming does. In most applications; it obtains solutions by working backward from the end of the problem toward the beginning. But, in solving the Niger- Delta problem of under-development, forward recursive approach is advanced.

Dynamic programming is a useful mathematical technique for making a sequence of interrelated decision. It provides a systematic procedure for the optimal combination of decisions. In contrast to linear programming, there does not exist a standard mathematical formulation of Dynamic programming problem. Rather, dynamic programming is a general type of approach to problems solving and the particular equation used must be developed to fit each situation. Therefore, a degree of ingenuity and insight into the general structure of dynamic is required to recognize when and how problem can be solved by dynamic procedures. The abilities can best be developed by an exposure to a variety of dynamic programming applications and a study of the characteristics that are common to all these situations (Fredrick et al , 2007).

Characteristics of Dynamic Programming Problem

i. The problem can be divided into stages, with policy decision required at each stage.

ii. Each stage had a number of stages associated with the beginning of that stage.

iii The effect of the policy decision at each stage is to transform the current stage to a state associated with the beginning of the next stage.

iv. The solution procedure is designed to find an optimal policy for the overall problem (i.e. a prescription of the possible states).

iv. Given the current state, an optimal policy for the remaining stages is independent of the policy decision adopted in previous stages.

v. The solution procedure begins by finding optimal policy for the last stage.

vi. A recursive relationship that identifies optimal policy for stage n given the optimal policy for stage n+l is available.

Recursive Nature of Computation in DP an Algorithm

Computations in dynamic programming are done recursively, so that optimum solution of one sub-problem is used as an input to the next sub-problem. By the time, the last sub-problem is solved; the optimum solution for the entire problem is at hand. Both forward and backward recursions yield the same solution. Although the forward procedure appears more logical, DP literature invariably uses backward recursion. The reason for this preference is that, in general backward recursion may be more efficient computationally. But, I do not think so. It rather makes it complicated. The study therefore, adopted the forward recursive analysis.

The procedure for solving a problem by using the dynamic programming approach can be summarized in the following steps;

1. Identify the problem decision variable and specify objective function to be optimized under certain limitations, (if any).
2. Decomposition of (or divide) the given problem into a number of smaller sub-problems (or stages). Identify the state variables at each stage and write down the transformation function as a function of the variable and decision variable at the next stage.
3. Write down a general recursive relationship for computing the optimal policy. Decide whether to use the forward or the backward method to solve the problem.
4. Construct appropriate tables to show the required values of the return function at each stage as one optimal policy.

Decision , d_n	$f_n(S_n, - d_n)$	**Optimal Return**	**Optimal Decision**
States S_{ndn}	$f_n^*(s_n)$	d_n	

The recursive relationship equation

$F_n^*(S_n) = \min d_n \{ D_{sn}, d_n + f_n - 1(d_n) \}$ n=1,2,3,4...

Where;

d_n = decision variables that define the immediate destination when there are n = 1,2,3,4 stage to go.

S_n = State variables describe a specific situation at any stage.

D_{sn} 'd_n= distance associated with the state variable, d_n for nth stage.

$f_n(S_n, d_n)$ = minimum total distance for the last n stage.

$f_n^*(s_n)$ = Optimal path (minimum distance)

$f_n - 1'(d_n)$ = The Optimal distance for the previous state.

To make the above recursive relationship equation clearer and understandable; it simply means

Optimal Distance = Optimal current distance + Optimal previous distance.

Dynamic Programming and the Niger-Delta Situation

As officially demarcated by the Federal government of Nigeria in the year 2000, the Niger-Delta region comprises of nine states of Abia, AkwaIbom, Bayelsa, Cross-Rivers, Delta, Edo, Imo, Ondo and Rivers. This was based on the fact that, they are oil producing States confronted with similar problem of under-development due to the negligence on the part of government and oil companies failing to impact development in the region despite the oil wealth that accrues from the region.

The under-development in the region is characterized by

i. Lack of road networks and bridges to open and link the hinter lands to the cities.
ii. Lack of jobs for the able-bodied youths of the region
iii. Lack of energy, electricity to attract industries.
iv. Lack of well equipped educational institutions to train manpower for the highly skilled jobs e.g the oil and manufacturing companies.
v. Lack of adequate health care facilities.
vi. Lack of sincere leadership
vii. No fertile land for mechanized agriculture due to environmental degradation arising from oil companies activities.
viii. No potable clean water supply (but water everywhere). What a paradox?
ix. Peace and security are eluding.
x. And these had metamorphosed into poverty and disease resulting to militancy, agitation for secession and premature mortality.

The above enumerated problems are compounded because of one influences the other and is by itself influenced by others. Considering the complex nature of these problems and looking at the nine states in the Niger Delta and the dire clamour for speedy development for the region, one has to consider Dynamic programming approach as a workable integrated development option for the region.

These problems cannot be solved at a go; nor is it possible to think of proffering solutions and getting them solved simultaneously. The application of dynamic model enables such a complex problem to be solved by breaking it into series of smaller and tractable sub-problems. This is exactly what dynamic programming does.

With the DP approach, the developmental issues would be handled on a stage by stage method. This will enable the government and companies to execute a particular project for a period of time and move to the next pressing development need for the region. Dynamic programming tool will help to progressively tackle the developmental challenges of the region while yielding optimal result.

2.2 Empirical Studies

Dynamic programming model has been used in solving wide range of problems in large organizations. In this study, we are pinning down to two major areas which are tied to the problem of underdevelopment in the Niger Delta, since we want to maximize the available resources and expertise and minimize wastages to the barest level with the use of a model.

The salesman problem solved (without a recursive equation).

The salesman wanted to maximize profit and time while minimizing cost of transportation. In Gupta and Hira (2005), the illustration of the salesman planning a business tour from Mumbai to Kolkata in the course of which purposes to cover one city from each of the company's different marketing zones enroute. As he has limited time at his disposal, he has to complete his tour in the shortest possible time. He drew a network which helped him to determine the optimum tour plan.

In his solution, there was no recursive equation and objective function stated mathematically. But, he succeeded in dividing the network into stages and analyzed additively to determine the optimal path way.

The minimal path solved (with a recursive equation)

Another similar illustration in Sharma, (2005) on short route problem of a salesman whose location in city A, decided to travel to city B, He knew the distance of alternative routes from city A to city B. He then drew a high network map showing the alternative route to analyze in order to arrive at the shortest route that covers the selected citied cities from A to B.

In this case, the salesman designed a recursive equation and adopted the backward approach. With the recursive equation, he obtained an optimal result- the minimal path.

From the forgoing empirical analysis, we can solve complex problems by first decomposing them into stages. The problem can further be tackled, depending on its nature with or without recursive equation as seen from the two illustrations. That is to say, problems may not be solved in the same manner, but ones aim is to achieve the optimum solution. The knowledge of tackling different problems from different angles is the ingenuity required in solving dynamic programming related cases like this underdevelopment problem in the Niger Delta.

The peculiar nature of the problem will determine the model to be developed. The sure way is that, the problem can be decomposed being a multistage problem. After decomposition, the recursive equation can be formulated to suit it (if need be).

2.3 Theoretical Framework

The research study is hinged on the theory of optimization or the principle of optimality. Dynamic programming is based on Bellman's principle of optimality which states that "An optimal policy (a sequence of decisions) has the property that whatever the initial state and decision are, the remaining decisions must constitute an optimal policy with regard to the state resulting from the first decision." This principle implies that a wrong decision taken at one stage does not prevent from taking optimum decisions for the remaining stages.

With this principle in mind, recursive equation is developed to take optimal decision at each stage. A recursive equation expresses subsequent state conditions and it is based on the fact that a policy is "optimal" if the decision made at each stage results in overall optimality over all stages and not only for the current stage.Dynamic programming as a mathematic technique deals with the optimization of multistage decision problems. The technique was initially called the stochastic linear programming. Today, dynamic programming has been developed as a mathematical technique to solve a wide range of decision problems and it forms an important part of every operation researcher's tool kit (Gupta and Hira, 1976).

CHAPTER THREE

RESEARCH METHOD

3.1 Research Design

Research designs are of three main types namely, survey design, experimental design and expost - facto design (ICAN PACK 2006). The study adopted both the survey and quasi-experimental design considering the nature of the research. The Niger-Delta problem of underdevelopment is a real life issue currently traumatizing the entire country.

Furthermore, the cross-sectional survey was used because it involves selecting samples of element for population of interest which are measured at a single point in time. The quasi- experimental design was adopted because the various elements of the design were not under the control of the researcher.

3.2 Population of the Study

The target population is the nine states in the Niger-Delta. Since the research problem is a sensitive one which demands intellectuals and professionals to respond to its questions, the target population was restricted to senior staff of the ministries of Economic Planning, Statistics and Strategy of the nine States. Moreso, the nine states are confronted, with virtually similar problems of land degradation, and other attendant sufferings from oil exploration and exploitation occasioned by neglect from oil companies and the Federal government. The research population also included staff of the Niger Delta Development Commission, NDDC (Port Harcourt), DESOPADEC (Warri) because of the nature of their commitment and services to the region.

3.3 Sampling Procedure

The technique adopted is the simple random sampling technique whereby every state in the population has an equal probable chance of being selected. This was achieved through the balloting or lottery method.

Sample

With the sampling techniques applied, Delta, River and Bayelsa State were selected. That means, data were collected from the ministries of planning of these States cum Niger Delta Development Commission, NDDC (Port Harcourt) and Delta State Oil Producing Area Development Commission, DESOPADEC, Warri, considering the nature of their work and relevance to the study.

3.4 Method of Data Collection

The research design resulted in the collection of secondary data from ministries of Economic planning and strategy of Delta, Bayelsa and River States and NDDC and DESOPADEC. Personal interview was also conducted with senior staff of Niger-Delta development commission,(NDDC) and Delta State Oil producing Areas Development , Commission (DESOPADEC, Warn). In the same vein, secondary data were sought from the data bank of the selected ministries and commissions, articles, journals, newspapers, and magazines as well as textbooks. This information helped in revealing critical points that are germane to the development of the Niger-Delta region. Specifically, secondary data inputted into the network model for analysis as regards the time frame of selected projects completed were sought directly from the ministries and commissions' data base to achieve objective of the study. .

3.5 Model Specification

Before designing the recursive equation for the model, let consider the major variables in the study. The Integrated plan and development of the Niger-Delta are the independent and dependent variable respectively. The holistic plan embodies the resources, and the expertise required. Development here depends on the integrated plan. That means, development cannot manifest without a plan.

Mathematically stated, since development, d is a function of the plan, p

$d = f(p)$

Where

Dependent variable, d = Development

Independent variable, p = Integrated plan

In this case of examining the possibility of developing a model for the execution of projects that will enhance the development, d of the Niger Delta, any project executed is regarded as a developmental step denoted as di. The impact of the project, ds is the return denoted as Ii

If Ii (d_i) denoted the return (impact from its activity (i.e projects executed), then total return may be expressed as,

$I_i (d_1, d_2 — d_n) = I_1 (d_1) + I_2 (d_2) + \ldots I_n (d_n)$ ……………… (1)

Under the constraints of resources and expertise

$d = d_i + \ldots + d_n,\ d_i \geq 0,\ i = 1,2,3$…………..(2)

The problem is to maximize development;

If $f_n\ (d) = \max\ I_n \{ (d_1\ d_2 \ldots d_n \} = \max \{ (I_1 (d_1) + I_2 (d_2) + \ldots I_n (d_n) \}$ …………… (3)

Where:

f_n (d) = maximum impact or return from all projects. An expression connecting $f_n(d)$ and f_{n-1} (d) may now be obtained using the principle of optimality. If $\mathbf{d_n}$ is the total projects to be executed in nth activity such that, $0 \leq d_n \leq \mathbf{d}$. Where one project has been executed, n-1 projects will be remaining. Let, f_{n-1} $(\mathbf{d} - \mathbf{d_n})$ denotes the impact from n-1 projects. Then total impact of nth activities will be $I_n d_n + f_{n-1} (d - d_n)$. An optimal choice of d_n will maximize the above function, thus, the model may be expressed as

$f_n(d) = \max \{ I_n(d_n) + f_{n-1}(d - d_n) \}, n = 2,3 \ldots\ldots\ldots\ldots (4)$

Where,

$f_1(d_1) = I_1(d_1) \ldots\ldots\ldots\ldots\ldots\ldots\ldots\ldots\ldots\ldots\ldots\ldots\ldots\ldots\ldots\ldots\ldots (5)$

and $f_1(d)$ = Optimal return

$I_1(d_1)$ = impact from project.

Therefore, in analyzing the model, the recursively equation to be used is;

$f_n(d) = \max, d_n \{I_n(d_n) + f_{n-1}(d - d_n)\}, \quad n = 2,3 \ldots\ldots\ldots\ldots (6)$

subject to;

$d_n \geq$ available resources

and

$d_n \geq$ expertise required

Where,

$f_n(d)$ = optimal impact (return from projects)

$f_{n-1}(d - d_n)$ = optimal for the previous projects

$I_n(d_n)$ = the optimal impact for the current project

d_n = state variable that determine the immediate project, where n = 1, 2,3,...

Conditions for the utilization of the model

Bear in mind that the impact or return from projects executed will be assessed

1. By the time frame of project completion.
2. By the cost of projects, where some relevant factors are assumed constant.

In addition, where is the impact of a project whose completion is deliberately prolonged? When the gains of developmental projects are maximized, the wastages that hitherto hinder the early completion of projects to specification and associated costs will diminish. Examples of such wastages are time due to abandonment of projects, inflated cost of projects due to poor planning and corruption in public offices.

We should re-collect that dynamic models help to model situation in which time (duration of project completion) often plays an important role.

3.6 Recursive Analysis by a Network Model.

Recursive analysis was the method adopted for data analysis.

Before, developing the model, here are the basic indicators of development (which would be incorporated into the model) as specified by the World Bank.

i. Adequate and nutritious food for every family.
ii. Portable drinking water for healthy living.
iii. Affordable houses (not promises).
iv. Standard affordable and accessible education for all with zero tolerance of illiteracy.
v. Affordable health care services.
vi. Constant supply of energy for sustainable economy activities.
vii. Long life expectancy.
viii. Employment Opportunities

These are just basic necessities of a decent life for every citizen not for a few according to Adebimpe. O.Ike (Daily sun, March 29, 200 pg 36). In addition to the above indicators of development, good network of road across the cities, towns and villages to ease transportation and facilitate other economic activities are very essential.

Network Model Design

The modelbelow was analyzed with the recursive equation formulated for it under model specification.

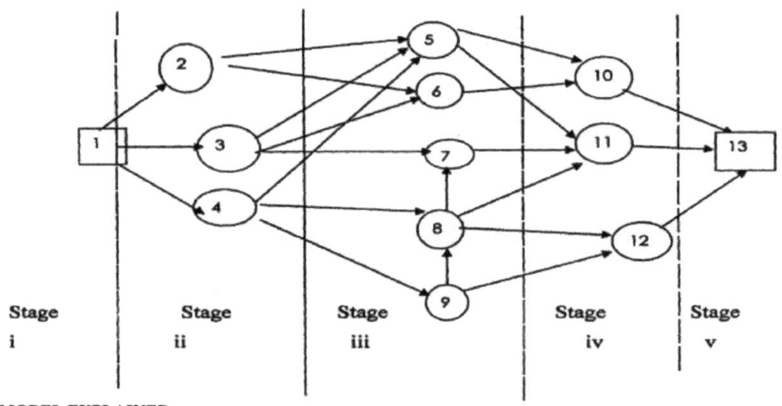

MODEL EXPLAINED

Figure 3.1: Designed Network for Analysis

The above network model represents the hypothetical journey from underdevelopment to development in the Niger-Delta in a forward and backward manner.

a) The rectangular nodes of 1 and 13 represent underdevelopment and development or vice visa. That is, whichever way one tackles the problem through the model, development will be achieved.

b) The circular nodes represent the indicators or parameters of development. Some parameters like corruption, politics and the likes hamper development.

c) The arrows indicate the duration of project completion in stages and possibly the cost. The model is divided into five stages.

d) From node 1 to 13 are different stages of decision taken (making) that will be implemented by government before attaining a reasonable and desirable level of development.

e) The model shows that, development is a gradual process and is from stage to stage. It is rather dynamic.

f) Again, development is a continuous one. It has no end

g) It should therefore be noted that, any particular level an economy is, there is a form of development (though it may not be appreciable). Therefore, continue to press on for more development until an internationally comparable standard is attained like the Developed world (Rodney, 1972).

h) It is worthy of note that, since we are finding the optimal pathway to development and that the forward Recursive Approach has been adopted, it is therefore adopted that node 1 is under- development stage in the study. We are entering some levels of development until a desirable level of development is reached probably at node 13.

i) The circular nodes from 2 to 12 represent the following development indicators (projects) and likely obstacles to development.

Node 2 - Militancy Eradication | Node 3 - Massive Agriculture

Node 4 - Road Networks and bridges | Node 5 - Portable Water Supply

Node 6 - Corruption Elimination | Node 7 - Education

Node 8 - Politicking Reduction | Node 9 - Energy (Electricity Supply)

Node 10 - Health Care Services | Node 11 - Employment

Node 12 - Peace and Security

Assumptioninthe Model

a) There would be political will supported by successive governments. That is, this programme must not be abandoned by government of the day.

b) There would be no interruption or break down of the process of activity.

c) There would be commission assigned to carry out this development programmes

d) There would be no manipulation or deviation from the original plan (except for minor changes).

e) There would be resources and expertise to execute the programmes.

CHAPTER FOUR

ANALYSIS OF DATA AND DISCUSSIONS

4.1 Model Analysis

The model was analyzed with the secondary data collected from the selected ministries and commissions of the respective states in order to achieve the purpose of the study. .

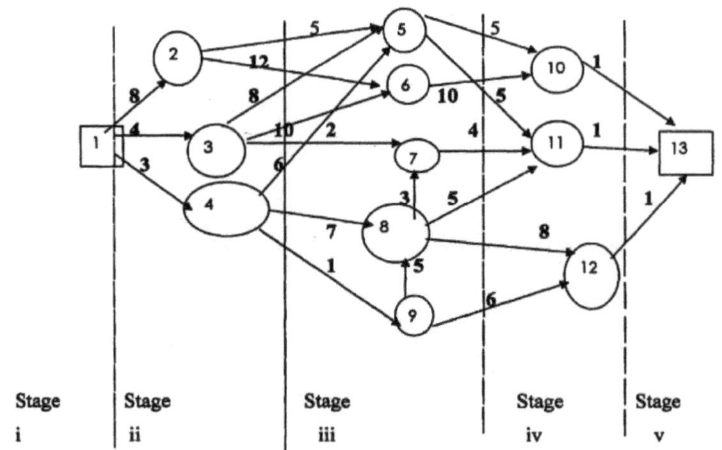

Figure:4.1 Formulated Network for Analysis.

Note: The average completion period (Time) of the selected and completed projects were imputed into the model for its analysis (see secondary data)

Explanations on the model

The circular nodes from 2 to 12 represent the indicators (projects) and obstacles to development have been explained in chapter three.

While the rectangular nodes of 1 and 13 represent underdevelopment and development since the forward recursive method has been adopted in this study. We should bear in mind that achieving development is not a free course. There are always notable impediments like politics, corruptions, natural disasters, world oil market dynamics etc which must be brought to fore when initiating a development plan in Nigeria.

Workings on the Model and Hints

1. Recursive formula for solving the model;

$$f_n (d) = \max d \{I_n (d) + f_{n-1} (d-d_n)\}, \qquad n = 2,3 \ldots$$

where,

$f_n (d)$ = optimal impact (return from projects)

$f_{n-1} (d- d_n)$ = optimal for the previous projects

$I_n (n)$ = the optimal impact for the current project

d_n= state variable that determine the immediate project,

2. We want to maximize development in the region by minimizing obstacles to accelerated development. Therefore in our calculations, we are minimizing time (Period) of project completion thereby reducing inflation and related matters. In short the workings dwelt on time or period of completed projects

3. Working forward in stages from node 1 to 13

for states,

S_1 =2, 3, and in stages 2, i.e. 3 projects in stage 2

S_2 = 5,6,7,8 and 9 in stage 3, i.e. 5 project in stage 3

S_3= 10, 11, and 12 in stage 4 , i.e. 3 projects in stage 4.

Working 1 on stage 3

(1). Starting with state $S_2 = 5$ (Project 5)

$f_2(5) = \text{Min}, (d_2)\ 2,3,4 \quad \left\{ \begin{array}{l} D_2'5 + f_1^*(2) = 5 + 8 = 13 \\ D_3'\ 5 + f_1^*(3) = 8 + 4 = 12 \end{array} \right\}$

$D_4'5 + f_1^*(4) = 8 + 3 = 11$

Decision = 11(S_1 =4)

That's project 4, Road Network is Chosen.

(2) Suite (S_2) = 6

$f_2\ (6) = \min d_2\ 2,3 \quad \left\{ \begin{array}{l} D_2'\ 6 + f_1^*(2) = 12+8 = 22 \\ D_2\ `+f_2^*(3) = 10 + 4 = 14 \end{array} \right\}$

Decision =14 (S_2=3)

That's project 3, food production (Massive Agriculture) was chosen.

(3). State (S_2) = 7

$f_2(7) = \min (d_2)3,8, \quad \left\{ D_3'\ 7 + f_1^*(3) = 2+4 = 6 \right\}$

Decision = 6 (S_1=3)

That's project 3, food production (Massive Agriculture) was chosen.

(4). State, $S_2 = 8 f_2(8) = \min (d_4),8 \left\{ D_4 8 + f_1^*\ 4 = 7 + 3 = 10 \right\}$

Decision =10 (S_1=4)

That's project 4, road network was chosen

(5) State (S_2) = 9

$f_2(9) = \min d_{4,9} \left\{ d_4,\ 9 + f1\ *\ (4) \quad = \quad 3 + 1 = 4 \right\}$

Decision = 4 (S_1 = 4)

That is project 4 , road network is chosen.
Results shown in the table

Decision	**F_2 (d)=max d_2 ($1_2(d_2)+f_{2\#}\ d_2$)**			**Min duration**	**Optimal Decision d_2**
States S_2	**2**	**3**	**4**	**$f2^*(S_2)$**	
5	13	12	11	11	4
6	22	14	-	14	3
7	-	6	-	6	3
8	-	-	10	10	4
9	-	-	4	4	4

<u>Working 2 on stage 4</u>
States or projects , S_3 in stage 4.
(6) State , $S_3 = 10$

$$f_3(10) = \min d_{3,}\ 5{,}6 \quad D_5, \begin{Bmatrix} 10 + f_2{}^*\ (5) = 5 + 11 = 16 \\ D_6,\ 10 + f_2{}^*\ (6) = 6 + 14 = 20 \end{Bmatrix}$$

Decision = 16 (S_2 = 5)

That's project 5 , portable water supply was chosen

(7) State , $S_3 = 11$

$$f_3(11) = \min d_{3,}\ 5{,}7,\ 8 \begin{Bmatrix} D_5,\ 11 + f_2{}^*\ 5 = 5 + 11 = 16 \\ D_7,\ 17 + f_2{}^*\ 7 = 4 + 6 = 10 \end{Bmatrix}$$

$$D_8,\ 11 + f_2{}^*\ 8 = 5 + 10 = 15$$

Decision = 10 (S_2 = 7)

That's project 7, education (eradicating illiteracy) was chosen

(7) State , $S_3 = 12$

$$f_3(12) = \min d_{3,}\ 8,\ 9 \begin{Bmatrix} D_{8,}\ 12 + f_2{}^*\ 8 = 8 + 10 = 18 \\ D_9,\ 12 + f_2{}^*\ 9 = 6 + 4 = 10 \end{Bmatrix}$$

Decision = 10 (S_2 = 9)

That's project 9, electrification (Energy) was chosen

Results shown in the table

Decision	**F_2 (d)=max d_2 ($1_2(d_2)+f_{2\#}\ d_2$)**					**Min duration $f_2(S_2)$**	**Optimal**
States S_2	**5**	**6**	**7**	**8**	**9**		**Decision d_2**
10	16	20	-	-	-	16	5
11	16	-	10	15	-	10	7
12	-	-	-	18	10	10	9

Working 3 Stage 5

(9) State , $S_4 = 13$

$$F_4(13) = \min d_4,\ 10, 11, 12 \quad \begin{Bmatrix} D_{10},\ 13 + f_3 * 10 = 1 + 16 = 17 \\ D_{11},\ 13 + f_3 * 11 = 1 + 10 = 11 \\ D_{12},\ 13 + f_3 * 12 = 1 + 10 = 11 \end{Bmatrix}$$

That is project 11 and 12 (employment generation and sustaining peace and security) were chosen

These results are shown in the table

Decision d_4	**f_4 (d) =d_4 1_4 (d_4) + f_4* d_4**			**Min duration**	**Optimal Decision**
State S_3	10	11	12	f_4*d_4	d_4
13	17	11	11	11	11 or 12

The above optimal result at various states and stages can be summarized as given below.

Entering Nodes or projects

(A) 1 ⟶ 4 ⟶ 9 ⟶ 12 ⟶ 13

(B) 1 ⟶ 3 ⟶ 7 ⟶ 11 ⟶ 13

Duration taken

(A) 3 + 1 + 6 + 1 = 11 yrs

(B) 4 + 2 + 4 +1 = 11 yrs

Alternatively, these states in their respective stages refer to projects executed.

From the above solutions, it is clear that, there are two alternative routes or pathway to development with minimum duration of 11years. This minimum duration was achieved through the decomposition of the underdevelopment problem into five stages by applying forward recursive relationship equation. This enabled us to select the best projects in accordance with priorities and developmental needs of the people, as shown below:

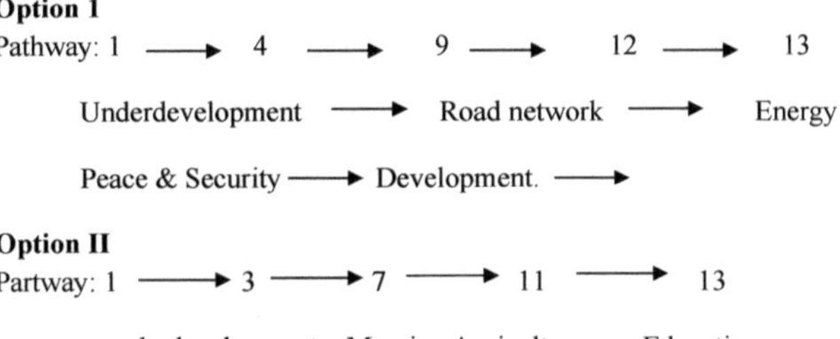

Option 1

Pathway: 1 ⟶ 4 ⟶ 9 ⟶ 12 ⟶ 13

Underdevelopment ⟶ Road network ⟶ Energy

Peace & Security ⟶ Development. ⟶

Option II

Partway: 1 ⟶ 3 ⟶ 7 ⟶ 11 ⟶ 13

underdevelopment , Massive Agriculture , Education,

, Employment , Development.

From the model, these other scenarios were examined without considering major planning factors like time, skill, money e.t.c

Entering Nodes

(a) 1	2	6	10	13
(b) 1	2	5	11	13

And Duration Taken Respectively

(c) 8	12	10	1 =	3lyrs
(d) 8	5	5	1 =	19yrs

These pathways constitutes longer years (Nigeria at 50, still grappling with basic development challenges), resulting to waste of resources, waste of time and energies causing confusion, resulting to rebellion and insurrection, debarring development; increasing poverty and diseases, causing avoidable deaths (kidnappings and assassinations), unnecessary deprivations, agitations for States creation, resulting to anarchy, secession campaigns and unending wars. These avoidable consequences were as a result of bad leadership, corruption, dirty politics and lack of long term planning in the polity.

Ordinarily, through any pathway, one can get to any desirable level of development but with grave consequences and development even eluding.

So, mere sitting of projects and political permutations in the form of appointment to ethnic nationalities crying of marginalization to douse social tension are no concrete plans to effect the type of development desired in Niger- Delta region.

IV. **Testing the Model and it Solution**

As already established in this study, a model is never a perfect representation of reality. But if properly formulated and correctly manipulated, it may be useful in predicting the effect of changes in control variable on the overall system effectiveness. The usefulness of a model is tested by determining how well it predicts the effect of these changes. Such an analysis is usually called **sensitivity analysis**.

The utility or validity of the solution can be checked by comparing the results obtained without applying the solution with the result obtained when it is used. To ensure reliability of the solution derived from the model, control must be established to indicate the limit the model and its solution is considered reliable.

It is needful to reiterate that, a model is the device which treats problem as a whole. It is essentially a hypothesis. Tentative explanation when formulated as preposition are called hypothesis. And that, hypothesis need not to be proved for every possibility in order to be acceptable. Sampling methods are usually sufficient to verify it. (Gupta and Hira, 2009).

The data required to carry out a test on the model are predictions made from its solution and data got from its implementation for the period taken. That's why the test is Post- Optimality Test. Bearing this in mind, the test for the model therefore needs time after its implementation by relevant bodies.

4.2 Discussions on Findings of the Study.

It was discovered that, some measures and tactics were initiated by successive governments for the development of the Niger-Delta. Though, these policies and measures were politically inspired, they lacked formidable political will required to implement them.

It was equally discovered that, in order to douse tension or agitations in the region, successive military governments created States to facilitate development, yet the region was enmeshed in abject poverty and underdevelopment. This is why the underdevelopment in the Niger-Delta is undoubtedly and unmistakably attributed to lack of purposeful leadership. In short, Niger-Delta problem is leadership problem according to Professor Chinua Achebe.

Another remote cause of underdevelopment of the region was the long military rule without germane, concrete and implementable plan for the advancement of the region in all ramifications. The region's upheavals and dire clamour for recognition resulting to militancy and kidnappings of expatriates was also fueled by the Multi-nationals (Oil Companies) who reneged on implementing their memorandum of understanding (MOU) agreement with host communities in the Niger-Delta. They exploit the region while the people (owners of the soil and mineral resources embedded in it) wallow in squalor disease, poverty and environmental degradation.

The Niger-Delta States compared to other parts of the country which are sustained by the wealth gotten from her land like Abuja, a dry land could developed very fast to a World-class capital city is an eye opener to the youths from the region which further strengthens the insincerity of government. This vented out the deep-seated anger of deprivation that past governments were playing the Niger-Delta people pranks. That is to say, the lack of development of the region was a deliberate ploy by the different governments since independence. And that all former and present commissions and ministries are only set-up to douse frictions, and to extend the hope for real development of the region.

It should be brought to the fore that, the various governments failed to nip the problem of under-development in the Niger-Delta in the bud. They (governments) felt adamant and callous until strikes are embarked on or oil pipelines have been exploded with powerful explosives (dynamites) by die-hard militants.

Another pertinent discovery in the course of this research was that, if concrete development programme is not rapidly embark on for the Niger- Delta region, the country may experience tsunamis' because, the anger seated in the minds of the youth is contagious and spreading very fast like wild fire. This is why the recommendation(s) in this project is/are sacrosanct for the country/region to develop evenly and have sustainable peace.

Summarily, other relevant findings were itemized as follows;

1. The region or country could be developed through a phase by phase development plan as show by the dynamic programming model.
2. Under development breeds hunger, poverty, sickness and diseases, corruption, kidnappings, militancy, deep animosity, assassination and even war.
3. Actually, our leaders grossly failed in executing development policies for the whole 43 years or more of Oil exploration and exploitation in the Niger-Delta. In fact, there was no developmental frame -work for the region's advancement.

CHAPTER FIVE

SUMMARY, CONCLUSION AND RECOMMENDATIONS

5.1 Summary

The development of the Niger-Delta region is not something the Federal and State governments should toil with any longer. Development is practical. It is measurable and manifestable in real terms. The Federal and State governments should be serious and sincere for once, and face the challenge squarely. The governments cannot afford to fail again, because the time is ripe for harvest, the resources are available and we can import the required technologies and expatriates from advanced nations in the world to speed up the development of the region/country.

Based on this backdrop, a Dynamic Programming Approach has been proffered in this research as a mathematical and scientific solution to solving the imbroglio of under-development in the Niger- Delta region. This Approach models a progressive and phase by phase development agenda for the region. The programme (model) will be more understood by experts (though simplified in the work) Therefore, such experts in the field of development be hired to handle the implementation of the model.

In a nutshell, the term development could be likened to a process that is progressive, evolving and technical (i.e. requiring technical know-how). In this case, development and growth should not be interrupted. Interrupting the Niger-Delta case will take us back to stagnation and the word of under-development. Henry ford said, ‘Nothing is particularly hard if you divide it into small jobs' this axiom is really captured by this dynamic programming model.

5.2 Conclusion

The development of any dtate and country depends on several factors namely, human capital, infrastructures, governments, resources, technologies, political will, people's co-operation, peace and development framework (i.e. the dynamic programming model) etc. Development is not one day or a year exercise, but it encompasses a long period of time. In fact, development is a long term project. It is rather a continuous process. And the type of development the Niger-Delta region deserves or is yearning for will take over 30 years or more to attain when requisite factors are garnered into the development programme. This is the sole reason, why a mathematically structured development framework must be put in place for the advancement of the region for uninterrupted period.

The region wants rapid development. The Niger-Delta people are keenly ready to give necessary support to other stakeholders like the Federal and State governments and the Multi-national companies to galvanize their resources and technologies in line with the marshaled programme for developing the region. The emphasis here is that, all hands (all stakeholders) must be on deck to fasten the region's development since a model has been established in this research to enhance its development.

5.3 Recommendations

Based on the findings from the study, we recommend as follows:

1. The Federal, State governments and the Multi-national companies must fear and respect GOD. The Holy Scriptures admonished in Proverbs 1 verse 7 that, the fear of God is the beginning of wisdom. The Oil and gas deposit in the Niger- Delta is a blessing from God and not a curse.
2. The Federal, State Governments and the Multi- nationals must show sincerity and maturity to actually develop the Niger- Delta.

3. The Federal and State governments should henceforth evolve a dynamic programming approach (DPA), or an integrated regional development plan (IRDP) for the region's development.
4. The Niger-Delta region should have an unbroken development for the next 30 decades which is conscientiously pursued by the Federal, State governments and multi-national companies.
5. Lastly, developmental projects should be evenly spread among the nine Niger-Delta States to avoid internal regional crisis. This will help to maintain and sustain the peace in the region/country.

5.4 Contributions to Knowledge

The essence of dynamic programming model is to solve complex problems from different possible ways. It is never a straight jacketed technique for solving peculiar issues. It could be applied to decompose problems defiling ordinary or normal linear solution. We may be thinking that, before we can solve any problem using dynamic programming approach, an objective function and more so, a recursive equation must be formulated. It is not necessarily so. Every problem has its own peculiarities and intricacies. The Niger-Delta question of under- development has its own distinctive features. With the knowledge of dynamic programming technique, complex problems in modern times can be broken down to solvable sub- problems and achieving optimal results.

With this orientation and training in mind, complex issues should be tackled from diverse possible ways while aiming at a definite goal of optimal result within its limitations or constraints. In management, there is no single best way of solving organizational problems. Through the knowledge of dynamic programming approach, every problem has a solution.

5.5 Suggestions for further research

The world is a dynamic place and technologies are rapidly changing and advancing as well as research is a continuous process. With increase in knowledge and the desire to know more heightened by the computer age, this research work could be deepened to critical sectors- Health, Education, Power, Transportation etc of the Nigerian Economy.

REFERENCES

Anyanwu and Oaikhenem, and et al (1997) The Structure of the Nigerian Economy (1977) Joanee Publishers Ltd Onitsha, Nigerian.

Daily Sun March 29, 2010

Enos, G (2004), Planning Development, Edited by Arthur Hazeiwood in Developmentseries, cited in Leonard (2000) Mathouse

Financial Vanguard November 3, 2008

Gupta andHira. D (1976) Operation Research, S chand and company Ltd New Delhi.

Institute of Chartered Accountants (2006) Study packfor professional Examination II On Research Methodoligy in Social Sciences and Education. Uniben Press, Benin-City.

Jhingan, M.L (2008), Economic Development and Planning. Vani Educational books, India.

Keenleyside ,H (1976) Obstacles and Means in International Development in Ham ridge (edited) Dynamics of Development, cited in Leonard (2000) Malthouse Press Ltd Nig.

Kellinger F.N. (1964) cited by G.O Yomere et al (1999) Research Methodology in Social Sciences and Education. Uniben press Benin- City.

King James Version (1979), The Holy Bible, Holman Bible polishers, Nashville Tennese.

Leonard, O.O (2000), Economic Policy and Economic Development in English Speaking Africa.Maithouse Press Limited, Nigeria.

Lewis, W.A. (1954), cited by Jhingan M,L (2008) Advanced Economic Theory, Nisha Enterprise and New Delhi.

Meler, G (1976), Leading Issues in Development, 31(1 Edition, New York, Oxford University press.

Murray R. et al (2005), Theory and Problems of Statistics. Tata Mcgraw-Hill Publishing Limited New Delhi.

Nachmias, D. and Nachmlas,C. (1981), Research Methods in Social Sciences .New York, St. Matins press.

Oboreh (2008).Unpublished Lectures' Note on Dynamic Programming

Rodney, W (2005), How Europe Under-Development Africa.Panaf Publishing Inc. Abuja, Nigeria.

Seers, D (1969) The Meaning of Development, Eleventh World conference of Society for International Development, New Delhi

Selltlz, C. et at (1981), cited by Yomere, G. O. et al (1999) Research Methodology in the Social Sciences and Education.
Uniben press Benin-City.

Sharma J.K. (2005) Operations Research, Theory and Applications,Schand and company Ltd, New Delhi.

Simon, K (1967), Population and Economic Growth, Proceedings of American philosophical Society Vol 3No. 13.

Simon, K (1971), Modern Economic Growth, Findings and Reflections.cited in Leonard (2000) Mathouse Press ltd, Nig.

Suchumpeter J.A (1934), cited by Jhingan M.L (2008) Advanced Economic Theory. Nisha Enterprises, New Delhi.

Sunday Sun may 23, 2010

Taha, H (2007), Operations Research , An Introduction. Macmillan Publishing Co. New York.

Tuesday Vanguard October 28, 2008

.Tull and Hawkins (1976), cited by G.O Yomere et al (999). Research Methodology in Sciences Education.Uniben Press Benin-City.

Vanguard, Newspapers: (2008) October 14,

Vohra, N.D. (2007), Quantitative Techniques in Management 3rd Edition. Tata McGraw- Hill Publishing Co. Ltd, New Delhi.

World Bank (1991), The Challenges of Development; World Bank Report New York. Oxford University Press

Yomere, G.O. and Agbanifoh B.A (1999), Research Methodology in the Social Science and Education. Centrepiece Consultants Nigeria Ltd Benin-city

Zeller, R.J. an Cramines, E.G (1979)Cited by G.O Yomere et al (1999) Research Methodology in the Social Science and Education. Uniben Press Benin- City.

APPENDIX A:

Chart Indicating the Completion Period (Time) of Selected and Completed projects by the governments of Delta, River, Bayelsa States and NDDC, PH and Desopadec, Warn

Projects (nodes)	**Mins /Comm**	**Completion time (yrs)**	**Average time**
Node 2 militancy eradication	Delta Rivers Bayelsa NDDC DESOPADEC	10 10 8 7 6	8yrs
Node 3 massive agriculture	Delta Rivers Bayelsa NDDC DESOPADEC	4 5 5 5 4	4yrs
Node 4 min of 4km road	Delta Rivers Bayelsa NDDC DESOPADEC	3 4 4 2 2	3yrs
Node 5 water supply scheme	Delta Rivers Bayelsa NDDC DESOPADEC	5 6 6 4 4	5yrs
Node 6 corruption elimination	Delta Rivers Bayelsa NDDC DESOPADEC	12 12 10 10 2	8yrs
Node 7 education six classroom block	Delta Rivers Bayelsa NDDC DESOPADEC	1 2 3 2 2	2yrs
Node 8 politicking reduction	Delta Rivers Bayelsa NDDC DESOPADEC	8 8 8 6 5	7yrs
Node 9 energy (eletrication of 3km)	Delta Rivers Bayelsa NDDC DESOPADEC	1 1 1 1	1yrs

Node 10 health care services (standard hospital)	Delta Rivers Bayelsa NDDC DESOPADEC	4 5 6 5 5	5yrs
Node 11 employment generation	Delta Rivers Bayelsa NDDC DESOPADEC	5 4 5 5 4	4yrs
Node 12 peace and security	Delta Rivers Bayelsa NDDC DESOPADEC	8 8 8 5 4	6yrs

Source: Being secondary data collated from the ministries of planning of Delta, Bayelsa, Rivers, NDDC (Port Harcourt) and DESOPADEC (Warri) used for network model analysis in chapter 4